CW01500146

THE POCKET

Night Sky

Published in 2026
by Gemini Gift Books
Part of Gemini Books Group

Based in Woodbridge and London

Marine House, Tide Mill Way,
Woodbridge, Suffolk IP12 1AP
United Kingdom

www.geminibooks.com

Text and Design © 2026 Gemini Gift Books Ltd
Part of the Gemini Pockets series

Cover illustration: Shutterstock/gianni triggiani

ISBN 978-1-80247-308-7

A CIP catalogue record for this book is available from the British Library.

Manufacturer's EU Representative: Eurolink Compliance Limited, 25 Herbert
Place, Dublin, D02 AY86, Republic of Ireland. admin@eurolink-europe.ie

Printed in China

10 9 8 7 6 5 4 3 2 1

Picture Credits: 4: Shutterstock/Jamo Images; 7: Shutterstock/Romolo Tavani; 8:
Shutterstock/Simon's passion 4 Travel; 20, 40, 110: VectorStock/Perezka_Klo 48:
Shutterstock/Angkara saputra; 51: VectorStock/dagadu; 54–56, 58, 60–62, 64:
VectorStock/macrovector; 72: Shutterstock/Wasim_shaikh; 78–85: Shutterstock/
WillDora; 96: Shutterstock/arvitalyaart; 114: VectorStock/A-R-T-U-R

THE POCKET

Night Sky

CONTENTS

Introduction   06

CHAPTER ONE
The Stars   08

CHAPTER TWO
The Planets   48

CHAPTER THREE
The Moon   72

CHAPTER FOUR
Other Celestial Bodies &
Space Phenomena   96

Introduction

To some, the night sky is just a backdrop for a good night's sleep. But to others, it is a celestial landscape to explore, even if being observing it from millions of miles away.

Of course, a quick look upward will give you a glimpse of the basics – the stars, the Moon. But what else is out there?

We're here to help you chart the night sky and uncover its best assets – from constellations to comets, planets, asteroids, meteor showers, satellites and everything in between.

Get ready to see everything our solar system and the stars beyond have to offer.

Chapter One

THE STARS

Stargazing Made Simple

You might think you need an expensive set-up – such as a telescope – to enjoy a night under the stars. But the truth is, you can spot constellations and admire the night sky with nothing more than your eyes.

Experts recommend finding a spot away from streetlights and other sources of light pollution. Let your eyes adjust to the darkness for about 20 minutes; then look up! You'll be amazed at how many stars come clearly into view.

Stargazing Tips

* You can go stargazing at any time of year, but autumn and winter are ideal – night falls earlier, giving you more time under the stars!

* If you're heading out during colder months, wear warm clothes and bring a hot drink in a thermos to stay cozy.

* For more comfortable viewing, consider relaxing on a camping mattress or reclining in a lounge chair.

* Bring a torch (flashlight), but opt for one with a red light bulb – it's gentler on your eyes in the dark.

* A good pair of binoculars can enhance your experience by bringing distant stars and celestial features into sharper focus.

* Download a stargazing or astronomy app to help identify constellations, planets and stars across the night sky.

What are Constellations?

A single star can be a breathtaking sight – but a group of stars that forms a familiar shape, known as a constellation, brings a whole new layer of magic to the night sky.

Constellations differ from asterisms – the name given to any identified star pattern . While asterisms can be informal groupings, constellations are officially defined and mapped with specific boundaries that, when combined together, encompass the entire night sky.

Spotting constellations takes a little imagination. You'll need to mentally connect the individual stars to see the shape they form. Once you can tap into that creative side of your brain, you'll find that an evening of stargazing becomes a journey through myth, science and wonder.

Let's explore some of the most well-known constellations and asterisms that you can pinpoint from your own backyard.

Ursa Major

Visible from: Northern Hemisphere
Best seen: All year-round, with peak visibility
from November to February

Ursa Major, also known as the Great Bear, covers
about 3.1% of the sky, making it the third-largest
visible constellation. It's a great one to learn early
on because it contains one of the most
recognizable star patterns: the Plough (also
known as the Big Dipper in the US and Canada).

In fact, one of the easiest ways to spot Ursa
Major in the sky is by first locating the Plough/
Big Dipper, thanks to its bright, easily identifiable
stars. From there, the rest of the bear's shape
begin to emerge – its body extends outward, with
long legs that appear to trail behind.

The Plough/ The Big Dipper

Visible from: Northern Hemisphere
Best seen: All year-round, with peak
visibility in spring

If you've spotted Ursa Major, then you've
already found the Plough/Big Dipper. This
asterism is comprised of seven bright stars
that form the shape of a ladle, with a long
handle and a cup at the end.

This easy-to-find constellation is often used to
introduce stargazers to the skies of the Northern
Hemisphere. For example, if you picture an
imaginary line upward from the outer edge of the
ladle (the bowl of the Plough/Big Dipper), it will
point you straight to Polaris, the North Star!

Orion

Visible from: Northern and
Southern Hemisphere
Best seen: January to April

The shape of Orion is often described as a hunter
shooting a bow and arrow. The trick to spotting
it is to find the three bright stars that make up
the figure's belt: from there, it's easy to find the
rest of the constellation, which features a
handful of noteworthy stars.

One star to look out for is called Betelgeuse,
which glows with a distinctive reddish hue. It's
the 11th-brightest star visible from Earth and
marks the shoulder of Orion. Surprisingly, it is
not the brightest of Orion's stars – that title
belongs to Rigel, the 7th-brightest star, which
marks one of Orion's feet.

Centaurus

Visible from: Southern Hemisphere
Best seen: March to May

Centaurus is the ninth-largest constellation, stretching over 1,060 square degrees of the sky. Its bright stars outline the image of a centaur – a mythical creature that is half-human, half-horse.

Two of Centaurus' stars – Alpha Centauri and Beta Centauri – can help stargazers locate another famous constellation: the Southern Cross. Simply imagine a line from Alpha through Beta and extend it upward – the line will pass directly through the Southern Cross.

Alpha Centauri is the closest star system to the solar system. It encompasses Omega Centauri, the brightest cluster of stars visible from Earth.

Southern Cross (Crux)

Visible from: Southern Hemisphere
Been seen: All year-round

The Southern Cross, also known as the Crux, is the smallest visible constellation – but it's also one of the easiest to find in the Southern Hemisphere. Its four main stars shine so brightly that even novice stargazers can spot it with ease.

As with many constellations, you'll need a little imagination. The stars mark the ends of a vertical line and a horizontal line. Connect these points in your mind, and you'll see the shape of a cross emerge in the sky!

Ursa Minor/ Little Dipper

Visible from: Northern Hemisphere
Best seen: All year-round, with peak visibility in June

Some know it as Ursa Minor, meaning "Little Bear", while others see the shape of a ladle and call it the Little Dipper. Either way, this constellation is considered a smaller counterpart to Ursa Major.

What makes Ursa Minor especially noteworthy is that is contains Polaris, also known as the North Star. Polaris sit less than one degree from true north, making it a key tool for navigation. Its steady position and brightness makes it remarkably easy to spot in the night sky.

The North Star

The North Star, or Polaris, is one of the most famous stars in the night sky. Located almost directly above the Earth's North Pole, it appears to stay fixed in place while other stars seem to rotate round it.

Because of its position, Polaris has long been used for navigation – especially by sailors and travellers before the invention of compasses and GPS, to help them determine the direction of north.

Fun fact: Polaris is actually a triple star system made up of three stars orbiting one another. What looks like a single point of light is, in reality, a complex celestial trio!

Honorary Star Names

Stargazing is an ancient practice – astronomers have been naming stars since long ago, but without modern technology, they could only pinpoint the most visible stars. Today, bright stars like **Vega** and **Sirius** are still recognized by the individual names they received hundreds, or even thousands, of years ago.

Some stars are named after famous or noteworthy astronomers – a rare honour that recognizes a scientist's significant contribution to the field of astronomy.

Scientific Star Names

Not all stars have names that come from mythology or are named after astronomers. In fact, most stars are identified using one of three scientific systems that help keep track of them:

* A Flamsteed number, which identifies most visible stars in southern England – assigned by astronomer John Flamsteed

* A Greek or Latin letter, known as a Bayer designation, assigned by celestial cartographer Johann Bayer

* A catalogue number, given to lesser known or less visible stars – typically made up of a combination of letters and numbers to allow for easy identification and distinction.

The Colour of the Stars

As previously mentioned, Orion contains the red-tinted star Betelgeuse. In fact, if you look closely, you'll notice that all stars can vary slightly in colour.

A star's colour is linked to its surface temperature. As a star produces more gas, it has more fuel to burn – this increases its temperature, which in turn changes its colour.

The hottest stars in the universe can reach surface temperatures of up to 40,000°C (72,000°F). But which colour is the hottest?

You might be surprised...

Blue

When stars reach temperatures of at least 28,000°C (50,400°F) and contain ionized helium, they appear dark blue. These are the hottest stars in the sky – burning at incredible temperatures.

Star colours range from blue to red, with cooler stars appearing in warmer hues. Here's a breakdown of star colours from hottest to coolest:

* **Dark blue**: 28,000+°C (72,000°F)
* **Light blue**: 10,000–28,000°C (18,000–72,000°F)
* **White**: 7,500–10,000°C (13,500–18,000°F)
* **Light yellow**: 6,000–7,500 °C (10,800–13,500°F)
* **Yellow**: 5,000–6,000°C (9,000–10,800°F)
* **Orange**: 3,500–5,000°C (6,300–9,000°F)
* **Red**: 2,000–3,500°C (3,600–6,300°F)

Zodiac Constellations

As the Earth orbits the Sun, our view of the night sky changes throughout the year. During this orbit, the Sun appears to pass through a group of constellations known as the Zodiac. The position of the Sun at the time of your birth determines which of these constellations is associated with your Zodiac sign – a system originally developed for use in ancient astronomy and calendars.

Aries

Visible from: Northern Hemisphere
Best seen: December

Aries is the second smallest of all the Zodiac constellations, covering just 441.39 square degrees of the sky. Despite its size, the constellation was historically used as the starting point for mapping the stars in the sky.

Aries contains only six named stars, most of which are fairly dim with only a few bright stars that are visible to the naked eye.

Taurus

Visible from: Northern Hemisphere
Best seen: January

Taurus is the sixth-largest Zodiac constellation, covering 797.25 square degrees of the sky. Its collection of stars form the shape of a bull, making it one of the more recognizable constellations.

The brightest star in Taurus is Aldebaran, a fiery orange star located about 65 light-years from Earth. Its warm glow marks the bull's eye, appearing to be gazing upward Orion, the hunter, in a striking celestial alignment.

Taurus also contains two well-known star clusters – the Hyades and the Pleiades – which are visible to the naked eye.

Gemini

Visible from: Northern Hemisphere
Best seen: February

Next to Taurus along the Sun's path lies Gemini, known as the twins of the Zodiac. This constellation is fairly easy to recognize – it resembles two stick figures standing side by side with their arms round one another!

Gemini's two brightest stars, Castor and Pollux, represent the heads of the twins. Both are named after figures from Greek mythology, but they're also scientifically notable: Pollux is an orange giant star, while Castor is actually a complex multiple star system.

Gemini is one of the smaller constellations of the Zodiac – it ranks eighth in size and covers 513.76 square degrees of the sky.

Cancer

Visible from: Northern Hemisphere
Best seen: March

Slightly smaller than Gemini, Cancer is the ninth-largest Zodiac constellation, covering 505.87 square degrees of the sky.

At first glance, Cancer can be tricky to spot, as it doesn't have many bright stars. But if you look closely, you might notice a faint V-shaped pattern that resembles a pair of pincers – giving a clue as to why this constellation is associated with a crab.

Cancer also contains a notable star cluster called Praesepe, or the Beehive Cluster, which is visible with binoculars or sometimes even to the naked eye under dark skies.

Leo

Visible from: Northern Hemisphere
Best seen: April

Leo is the third-largest constellation in the Zodiac, covering a large section of the night sky. It's traditionally associated with the shape of a lion, and its outline is one of the easier Zodiac figures to recognize.

You can easily spot Leo by its brightest star, Regulus. This shining light marks the base of the Sickle – a curved set of six stars that makes up the lion's head and mane. Regulus is a blue-white star that is part of a multiple-star system.

Virgo

Visible from: Southern Hemisphere
Best seen: May

Virgo is not only the largest Zodiac constellation, but also the second-largest constellation in the entire night sky, spanning an impressive 1,294.43 square degrees.

To find Virgo, start by pinpointing its brightest star, Spica, which you can spot by following the curve of the Plough's handle.

Whether or not you see the figure of a maiden – the symbol traditionally associated with Virgo – is up to your imagination. But its size, brightness and location make it a key constellation in the night sky.

Libra

Visible from: Southern Hemisphere
Best seen: June

As the seventh-largest constellation in the Zodiac, Libra sits between Taurus and Gemini, covering 538.05 square degrees of the sky.

Located next to Scorpio, Libra has two main stars – Zubeneschamali and Zubenelgenubi – which were once considered part of the scorpion's claws in ancient star maps. Today, they mark the outline of the scales, the symbol of balance that Libra is best known for.

Scorpio (Scorpius)

Visible from: Southern Hemisphere
Best seen: July

Despite being the third-smallest Zodiac constellation at 496.78 square degrees, Scorpio is one of the easiest to identify in the night sky. Its distinctive shape resembles a scorpion, with curved pincers and a long, looping tail.

Scorpio contains Antares, a red supergiant star that marks the heart of the scorpion and is one of the brightest stars in the sky.

To trace the rest of the constellation, start at Antares and trace the pattern down to find the scorpion's stinger.

Sagittarius

Visible from: Southern Hemisphere
Best seen: August

The fifth-largest Zodiac constellation, Sagittarius, covers a wide area of the sky and is represented by a centaur drawing a bow and arrow.

Although the full centaur shape can be hard to visualize, Sagittarius contains a well-known asterism called the Teapot, formed by eight of its brightest stars. This teapot shape is much easier to spot and serves as a helpful guide for locating the constellation.

The star at the teapot's spout also marks the tip of the archer's arrow – aimed directly at Scorpio!

Capricorn (Capricornus)

Visible from: Southern Hemisphere
Best seen: September

Capricorn is both the smallest and one of the most subtle constellations in Zodiac, making it a challenge to spot with the naked eye.

To locate this constellation, look east of the Teapot asterism in Sagittarius where you'll find a faint V-shaped pattern of stars that outlines the head and horns of the sea goat – a mythical creature with the front half of a goat and the tail of a fish.

Aquarius

Visible from: Southern Hemisphere
Best seen: October

Aquarius – the water-bearer – is the second-largest constellation, spanning 979.85 square degrees. Despite its size, it can be difficult to spot due to its relatively dim stars and scattered shape.

A helpful way to locate Aquarius is by finding the Square of Pegasus – a bright group of four stars in the northern sky. Just below this square is a Y-shaped pattern of stars, which represents the water jar!

Pisces

Visible from: Northern Hemisphere
Best seen: November

Pisces is the fourth largest Zodiac constellation, covering a wide section of the sky. It represents a pair of fish connected by a long line of stars and is located near the Square of Pegasus.

To locate Pisces, look just below the Square of Pegasus to find a small, round set of stars known as the Circlet – this marks the head of the western fish. From there, follow a V-shaped line of stars that stetches across the sky, leading to the eastern fish, which has a more triangular-shaped head.

"Do not look at stars as bright spots only. Try to take in the vastness of the universe."

American astronomer Maria Mitchell
(1818–1889)

1,000,000,000,000,000,000,000,000

That's one septillion – the estimated number of stars in the entire universe, according to astronomers.

The Milky Way – the galaxy in which our solar system exists – contributes more than 100 billion stars to that figure.

One

**That's how many stars are in our
solar system – only the Sun!**

The millions of other stars we can see are part
of the Milky Way, and many of them are likely to
have their own planets around them.
Considering there are an estimated 200 billion
galaxies in the entire universe... it's easy to
wonder what else might be out there.

Why Do Stars Twinkle?

When you look up at the night sky, you might notice that stars seem to twinkle – but they're not actually flickering. What you're seeing is caused by the Earth's atmosphere.

As starlight travels through the layers of air surrounding our planet, it is bent and scattered by pockets of air that are moving and shifting in temperature. This causes the light to change direction slightly by the time it reaches our eyes, making the star appear to shimmer or flicker.

Star Clusters

Not all stars live alone – some form in groups called star clusters.

A star cluster is a group of stars that were born from the same giant cloud of gas and dust. Because they formed together, the stars in a cluster often move through space as a group.

There are two main types of star clusters:

1. Open clusters are loose and contain a few hundred stars.
2. Globular clusters are tightly packed in a spherical shape – usually containing thousands to millions of stars.

Shooting Stars

You might have spotted something moving in the night sky during an evening of stargazing and thought it was a shooting star – but it's not actually a star at all.

What we call shooting stars are really meteors: tiny bits of space dust that enter the Earth's atmosphere at high speed. Most meteors burn out in the atmosphere before they make it to Earth's surface, creating a beautiful, glowing streak of light.

Sometimes, larger pieces survive the journey – these fallen shooting stars are called meteorites.

Light-year

Nothing moves faster than light! So, astronomers use a special unit of measurement to measure just how far it can travel: the light-year.

As the name implies, a light-year is the distance light travels in one year – about 5.88 trillion miles (9.46 trillion km). That's roughly 11 million miles per minute, or 186,000 miles per second.

For perspective, Earth is about 8 light-minutes from the Sun.

If you could travel at the speed of light, it would still take you 1.87 years to reach the edge of our solar system.

The closest-known star from Earth is...

The Sun

The star closest to Earth is the Sun – and nothing else even comes close.

The Sun is 93 million miles (150 million km) away from us. The next nearest star, Proxima Centauri, is a staggering 25 trillion miles away (about 40 trillion km). So, in terms of distance, the Sun has no real competition!

This is just one of many fascinating facts about the Sun!

Not only is the Sun the closest star to us –
with a diameter of about 865,000 miles
(1.4 million km) – it's also the largest object
in our solar system.

To put this into perspective, the Sun is
100 times wider
than Earth, and ten times wider than
the largest planet, Jupiter.

The Sun's massive dimensions create the
gravity that holds the rest of the solar
system in place – that includes everything
from planets and other stars to the tiniest
specs of space dust. Everything moves in
orbit round the Sun!

But beyond our solar system, our Sun
wouldn't make as much of an impact – it's
just an average-sized star compared to
others that astronomers have discovered.

The furthest-known star from Earth is...

Earendel

The Hubble Space Telescope first spotted
Earendel in early 2022, and NASA's James
Webb Telescope later explored further to
learn more about it.

Earendel is located a staggering 28 billion
light-years from Earth. The light we see
from it today was actually emitted about
12.9 billion years ago - making it the oldest
and most distant star ever observed.

The Lifetime of a Star

The Sun is vital to life on Earth – without its light and energy, humankind wouldn't survive. But like all stars, the Sun won't last forever – eventually, it will run out of the gas that fuels it!

Experts estimate that the Sun is about 4.5 billion years old and is roughly halfway through its lifespan. In roughly 5 billion years, it will swell into a red giant, eventually engulfing Mercury and Venus – and possibly Earth – before fading away.

Larger stars burn through their fuel much faster, so they have shorter lifespans, sometimes only a few million years. In contrast, smaller stars burn more slowly and shine less brightly, so they can live for trillions of years.

Chapter Two

THE PLANETS

The Solar System

There are eight planets within our solar system:

Mercury
36 million miles
(58 million km)
from the Sun

Venus
67.2 million miles
(108 million km)
from the Sun

Earth
93 million miles
(149.7 million km)
from the Sun

Mars
141.6 million miles
(227.9 million km)
from the Sun

Jupiter
483.7 million miles
(778 million km)
from the Sun

Saturn
889.8 million miles
(1.4 billion km)
from the Sun

Uranus
1.8 billion miles
(2.9 billion km)
from the Sun

Neptune
2.8 billion miles
(4.5 billion km)
from the Sun

Inner Planets

The first four planets **closest** to the Sun are:

* Mercury
* Venus
* Earth
* Mars

They are known as the Inner Planets. They're also called the terrestrial planets because they have solid, rocky surfaces made of rock and metal, unlike the gas giants further out in the solar system.

Outer Planets

The four planets **furthest** from the Sun are:

* Jupiter
* Saturn
* Uranus
* Neptune

They are known as the Outer Planets. Unlike the rocky Inner Planets, these don't have solid surfaces.

Jupiter and Saturn are known as gas giants, made mostly of hydrogen and helium that swirl round dense, rocky cores.

Uranus and Neptune are called ice giants. They're made of heavier elements like oxygen, nitrogen, carbon and sulphur – compounds that were frozen when the planets formed. These planets have slushy, icy cores surrounded by thick layers of swirling gas.

Mercury

Not only is Mercury the closest planet to the Sun, it is also the smallest planet in the solar system.

A year on Mercury – a single orbit round the Sun – takes about 88 Earth days. Because its axis is barely tilted, Mercury doesn't experience seasons like the Earth does; the amount of sunlight it receives remains consistent through the year.

Despite being so close to the Sun, Mercury has no atmosphere to trap heat. As a result, its surface reaches 430°C (800°F) during the day, but plunges to -180°C (-290°F) at night. This makes it both scorching hot and freezing cold!

And yet, Mercury isn't the hottest planet in the solar system...

Venus

Venus is often called "Earth's evil twin" – it's similar in size and structure to our planet, but far less friendly.

Its thick atmosphere traps heat through a powerful greenhouse effect, making Venus the hottest planet in the solar system. The surface reaches temperatures of around 475°C (900°F) – hot enough to melt lead!

Venus also spins in the opposite direction to Earth, so the Sun rises in the west and sets in the east. But you'd have to wait a while to see it – a single day on Venus lasts 117 Earth days.

Interestingly, about 30 miles (50 km) above Venus's metal-melting surface, temperatures are much milder – ranging between 30–70°C (86–158°F). Scientists believe this zone might be able to sustain tiny microbes... fancy a holiday above Venus?

Earth

Next in line is our home planet – Earth, a truly unique planet with the perfect conditions to support life.

Earth's atmosphere is comprised of 78 per cent nitrogen, 21 per cent oxygen and 1 per cent other gases. This special mix allows us to breathe and helps protects us from meteoroids by burning them up before they hit the surface.

It's also the only planet we know of with liquid water on its surface – and where life began around 3.8 billion years ago.

Earth is tilted at 23.4 degrees, and this tilt is what gives us seasons, as the Sun shines more directly on either the Northern or Southern Hemisphere during different times of the year.

Earth is only the fifth largest of all planets.

Imagine that the Sun is the size of a standard doorway – in that universe, the Earth would be the size of a coin in comparison!

Mars

Next up is Mars, which is about half the size of Earth. If Earth were the size of a coin, Mars would be the size of a raspberry!

A day on Mars is similar to ours, lasting 24.6 hours, but a year takes 687 days. Mars also has two moons: Phobos and Deimos. Phobos is slowly spiralling toward Mars, and experts predict that in about 50 million years it may crash into the planet or break apart, potentially forming a ring of dust and debris.

Mars is home to Olympus Mons, the largest known volcano in the solar system. It's nearly three times the size of Mount Everest, and its base would cover the entire US state of New Mexico.
Mars was once warmer and likely covered in water, which has led scientists to wonder if Mars could have supported life. So far, no signs of life have been found – but the search continues.

The surface of Mars is as cold as the South Pole – 60°C (140°F)!

Scientists have explored ways to warm up Mars – just in case we ever need to leave Earth. One idea is to install giant mirrors in space or on the planet's surface and reflect the Sun's light onto the planet. This would help raise the temperature and possibly make Mars more hospitable over time.

Jupiter

Meet the solar system's largest planet!

If Jupiter was a hollow shell, it could fit 1,000 Earths inside. In fact, it's twice as large as all the other planets combined.

Jupiter has the shortest day of any planet – just 10 hours long – but takes 12 Earth years to complete one orbit round the Sun.

Jupiter has a fascinating system of moons – four large moons and dozens of smaller ones. Europa, one of Jupiter's moons, contains a vast ocean beneath its icy surface – making it one of the best places to look for life beyond Earth.

Saturn

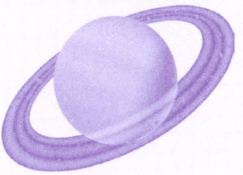

Saturn was discovered with the naked eye – it's the furthest planet from the Earth to have been first seen without a telescope or other device!

If you lined up nine Earths side by side, they would nearly stretch across Saturn's diameter – and that's not even counting its famous rings.

Saturn has 53 known moons and 29 more awaiting confirmation, meaning it could have a whopping 82 moons orbiting it!

The planet's most stunning feature is its system of seven rings, which appear solid from a distance but are actually made of billions of pieces of rock, ice and dust that float together in a ring-like orbit round the planet. Each ring orbits Saturn at a different speed, and together they stretch as far as 175,000 miles (282,000 km) from the planet's edge.

Uranus

Uranus stands out in the solar system because of its unusual tilt – its axis is tipped at nearly 90 degrees, making it appear to spin sideways – almost as if a ball was rolling round the Sun.

This extreme tilt causes some strange seasons on Uranus. One pole of the planet receives constant sunlight for 21 Earth years, while the other pole sits in complete darkness for the same amount of time.

Uranus is four times wider than Earth – if Earth were a coin, Uranus would be the size of an orange.

Uranus's Unique Moons

Uranus is also unique because its moons aren't named after characters in Roman or Greek mythology, like those of other planets. Instead, 27 of its 28 moons are named after characters from the works of Shakespeare and from Alexander Pope's poem 'The Rape of the Lock'.

1. Cordelia
2. Ophelia
3. Bianca
4. Cressida
5. Desdemona
6. Juliet
7. Portia
8. Rosalind
9. Cupid
10. Belinda
11. Perdita
12. Puck
13. Mab
14. Miranda
15. Ariel
16. Umbriel
17. Titania
18. Oberon
19. Francisco
20. Caliban
21. Stephano
22. Trinculo
23. Sycorax
24. Margaret
25. Propsero
26. Setebos
27. Ferdinand
28. S/2023 U 1 – this one hasn't gotten its literary name yet!

Neptune

Neptune is the only planet that's invisible to the naked eye, which isn't surprising considering it's more than 30 times further from the Sun than Earth is.

A single Neptunian year lasts 165 Earth years – this is how long it takes to make a single loop round the Sun. Though, like Earth, Neptune sits on a 28-degree tilt, meaning it experiences four seasons, but each one lasts about 40 years due to the planet's long orbit!

Neptune also has a faint ring system, with five known rings and four arcs – partial rings that don't form complete loops. It boasts 16 confirmed moons, though astronomers believe there could be more that have yet to be discovered!

The Planet Predicted by Mathematics

If Neptune is invisible to the naked eye, then how was it discovered?

The answer: *mathematics*

After the discovery of Uranus in 1781, scientists noticed something strange – its orbit didn't quite follow the rules of Newton's law of universal gravitation. So, they theorized that there could be another unseen planet beyond Uranus that was affecting its motion. Two astronomers – Jean-Joseph Le Verrier and John Couch Adams – independently calculated where this mystery planet might be.

In 1846, Le Verrier sent his predictions to German astronomer Johann Galle, who used a telescope to search the skies. Amazingly, Galle found the planet within one degree of Le Verrier's prediction.

The discovery of Neptune was a major breakthrough in celestial mechanics and the movements of objects in space – proving that careful mathematics could uncover worlds no one had ever seen.

What Happened to Pluto?

You might remember when our solar system had nine planets – including Pluto, which came after Neptune. However, in 2006, the International Astronomical Union (IAU) demoted Pluto from a fully fledged planet to a dwarf planet.

According to the IAU, a dwarf planet is an object that:

* orbits the Sun
* is large enough to pull itself into a nearly round shape
* has not cleared its orbit of other debris

In Pluto's case, its orbit overlaps with the Kuiper Belt – a region of icy objects beyond Neptune. That's what sets it apart from the eight official planets.

There are five dwarf planets within our solar system. They are:

1. **Ceres**, the smallest dwarf planet, sits between Mars and Jupiter and is the closest dwarf planet to the Sun.

2. **Pluto**, the largest dwarf planet, is one-fifth the size of Earth. It's the next planet after Neptune, 3.67 billion miles (5.9 billion km) from the Sun.

3. **Haumea**, which comes after Pluto, is one-seventh the size of the Earth and 4 billion miles (6.5 billion km) from the Sun.

4. **Makemake**, the next dwarf planet in the line-up, is 4.3 billion miles (6.9 billion km) from the Sun.

5. **Eris**, the second-largest dwarf planet, is furthest from the Sun – a staggering 6.3 billion miles (10 billion km) away from the biggest star in our solar system.

Spotting Other Planets from Earth

It's no secret that you can see stars and constellations, but did you know that you can also see some other planets by simply looking up at the night sky?

Planets don't produce their own light like stars. Instead, they reflect the Sun's light, and how bright they appear depends on:

* The size of the planet
* The reflectivity of its surface
* The amount of cloud coverage that deflects the glow

Based on these factors, the easiest planets to spot with the naked eye are:

* Mercury
* Venusw
* Mars
* Jupiter
* Saturn

Mercury

Mercury can sometimes be tricky to spot, despite that fact that it's close to the Sun. It sits low in the sky and is often hidden by sunlight.

Your best chance to see Mercury in the sky is about an hour before sunrise or after sunset, especially during spring evenings or autumn (fall) mornings.

Venus

Venus passes closer to the Sun than it does to the Earth, which explains why it's the brightest planet in the sky and can be seen with the naked eye.

You can see Venus for most of the year, but it's easiest to spot just before sunrise in the east or just after sunset in the west. The only time it disappears is when it passes behind the Sun for a month or two each year.

Mars

It can be so easy to see this fiery red planet with your naked eye that seeing it up close with a telescope can feel almost underwhelming!

Mars is visible from sunset until sunrise, as it moves in opposition to the Sun, rising when the Sun sets.

Jupiter

In June, Jupiter becomes the brightest celestial body in the sky – outshining even the stars! This happens as Venus disappears behind the Sun, leaving Jupiter to take centre stage.

Look for it glowing brightly in the western sky after sunset.

Saturn

Saturn is the furthest visible planet on this list, but it's surprisingly easy to spot if you know when to look! It's brightest from May to July when it remains visible after the Sun sets.

From late May to early June, Saturn is in opposition to the Sun, meaning it rises as the Sun sets. To find it, look up to the sky and search the space opposite the Sun – Saturn appears as a steady, golden glow.

Can I see Saturn's rings?

You can spot Saturn with the naked eye (a steady, golden-coloured dot in the sky) but not its famous rings. For that, you'll need a small telescope. Even at 25x magnification, the rings become visible. At 50x, you can see that the rings are separate from the planet's surface – a truly incredible sight!

Chapter Three

THE MOON

The Moon

You don't need help spotting the Moon – but it's far more than just a glowing object in the night sky. The Moon plays an important role in how our planet works.

Before the Moon settled into orbit, a day on Earth lasted just six hours. But once the Moon formed, its gravitational pull began to slow Earth's rotation.

As the moon gradually moved further away, Earth's spin continued to slow. Over hundreds of millions of years, this led to the 24-hour day we now experience.

Scientists also believe that the giant impact that caused the formation of our Moon also pushed Earth to tilt on its axis. Earth's 23.5-degree tilt is what gives us our seasons – and provides just enough Sun exposure to all corners of the globe.

Thanks to the Moon's gravitational pull, this tilt has remained stable for billions of years. Without the Moon, Earth might wobble unpredictably on its axis. That's why some scientists compare the Moon to training wheels – keeping our planet balanced as it spins through space.

The Moon and the Tides

The Moon also has a hand in creating waves in our oceans, seas and even our largest lakes.

That's because the Moon's gravitational pull causes the Earth's crust to lift slightly. As it orbits, this gravitational force creates a so-called "bulge" in the water, which pushes out to shore in the form of waves and tides.

The Changing Face of the Moon

Like Earth, the Moon has a day and night a side –
one half is always lit by the Sun, while the other
half remains in darkness. As the Moon rotates
through its orbit, different angles of its sunlit
side become visible to us.

This changing view causes the Moon to appear
in different shapes, or phases, throughout the
month. Sometimes we see just a sliver (a
crescent), and other times we see the entire
face (a full moon). These changes are called the
phases of the moon, and they follow a regular
29.5-day cycle.

New Moon

The name New Moon might sound like the
Moon should appear in the sky bright and
renewed, but it's actually the opposite. A New
Moon occurs when the illuminated side of the
Moon faces the Sun, while its darkened side –
the side experiencing night-time – faces Earth.
Because of this, the Moon appears invisible
to us from the ground. So during a New Moon,
there's no visible Moon in the sky – just
darkness where it normally shines.

Waxing Crescent

After the New Moon, the sunlit side of the Moon begins to shift back into view from Earth – but it takes time before we see it fully lit.

The first visible phase is the Waxing Crescent, when just a thin sliver of light appears along one edge of the Moon. This gentle glow grows a little more each night as the Moon continues its orbit.

First Quarter

It may look like half the moon is lit up, but this phase is called the First Quarter – and for good reason.

At this point, the Moon is a quarter of the way through its 29.5-day cycle. And while we see half of the Moon's face illuminated, remember: the Moon is a sphere. So what we're actually seeing is just a quarter of the entire surface.

Waxing Gibbous

As the month goes on, we see more and more of the Moon's illuminated side.

The Waxing Gibbous phase occurs when more than half of the Moon is lit up, but it's not yet full. At this stage, the majority the Moon's brightened face is visible from Earth, growing slightly larger each night.

Full Moon

Halfway through the Moon's cycle, we reach the Full Moon – when the entire sunlit side of the Moon is visible from Earth. This occurs when the Moon is positioned directly opposite the Sun, allowing us to see its full day side.

A Full Moon rises at sunset, shines all night and then sets at sunrise. While the phase itself doesn't last long, the Moon can appear full or bright for a few nights before it begins to wane and shift into the next phase.

Waning Gibbous

What waxes must wane – and after the Full Moon, the light begins to shrink. The Waning Gibbous phase is the first step in this process, wherein the illuminated side of the Moon starts to get smaller. During this phase, the Moon starts to rise later each evening, slowly shifting its position in the night sky as the cycle continues.

Last Quarter

Eventually, the Moon reaches its Last Quarter phase. Once again, half of the Moon's face is visible from Earth – but this time, it's the opposite half from what appears during the First Quarter.

Waning Crescent

The final phase of the lunar cycle is the Waning Crescent. At this point, the Moon has shrunk back to a sliver of light, barely visible in the early morning sky. This thin crescent means the Moon is nearly aligned with the Sun once more. Soon, it will face the Sun completely, becoming a New Moon, and the cycle begins once again.

Which Side of the Moon Are We Seeing – and When?

Trick question: we always see the same face of the Moon!

That's because the Moon takes just as long to rotate on its axis as does to orbit the Earth. This perfect timing means the same face of the Moon is always turned toward us.

This phenomenon is called synchronous tidal locking, and it's not unique to our Moon. Many large moons in the solar system do the same, always showing one face to their planet. Even some binary stars stay locked in this kind of cosmic rhythm.

It's All a Matter of Perspective

While stars and constellations appear (or don't appear) in certain parts of the globe, the Moon is visible everywhere. We can all see it, no matter where we are – and it's always in the same phase for everyone, no matter your vantage point. However, the appearance of the phase can be flipped depending on your location on Earth.

For example, if you're in the Northern Hemisphere and see a Waning Crescent on the left side of the Moon, someone on the Southern Hemisphere would be seeing the same phase, but the Waning Crescent would appear on the right.

The Daytime Moon, Explained

You might glance up during the day and notice something strange in the sky – the Moon!

Although it's usually associated with night, the Moon can be visible during the day several times a month. It often appears faint and pale, but it's definitely there.

The Moon is most visible during the First and Last Quarter phases, when it's high in the sky and the Sun hits it at a 90-degree angle. This makes the light bright enough to reflect off the Moon so that we can see it during the day.

In fact, the Moon can be seen during the day at any point in its cycle – except during the New Moon phase, when it's invisible, and during the Full Moon phase, when the Moon rises and sets.

Eclipses: When the Moon Becomes the Star

Sometimes, the Sun, Moon and Earth line up just right – and when they do, they cast a stellar shadow known as an eclipse.

There are two main types of eclipses:
* **Solar eclipse** – happens when the Moon passes between the Earth and the Sun, blocking sunlight and casting a shadow on Earth
* **Lunar eclipses** – happens when the Earth comes between the Sun and the Moon, blocking sunlight and casting a shadow on the Moon

Lunar Eclipses

A total lunar eclipse occurs when the Earth moves between the Sun and the Moon, blocking direct sunlight from reaching the Moon's surface. This casts a shadow over the Moon's surface – but it doesn't stay dark for long.

Some light still reaches the moon by bending round the edges of Earth. As the light passes through our atmosphere, shorter wavelengths like blue and violet are scattered. Only the longer red and orange wavelengths make it through – causing the Moon to glow a deep red or orange. This is why a total lunar eclipse is often called a "Blood Moon". If Earth's atmosphere is particularly dusty or cloudy when this happens, the fiery hue becomes even more pronounced.

During a partial lunar eclipse, only part of the moon enters the Earth's shadow. So, you won't see the reddish hue – it will stay more shadowed and grey.

Solar Eclipses

A solar eclipse occurs when the Moon passes between the Earth and the Sun, blocking some or all of the Sun's light from reaching up. While the Sun seems to take centre stage, the Moon is a major player in making this happen.

Solar eclipses can only happen during a New Moon, when the Moon's dark, unlit side faces Earth. The Moon lines up perfectly between the Sun and the Earth, blocking sunlight and casting a shadow to Earth.

When the Sun is fully blocked, only geographic areas within the darkest, central part of the Moon's shadow's (called the umbra) will see a total eclipse. If they're within the wider shadow (called the penumbra), they'll see a partial eclipse.

It only lasts for a few minutes, but a full solar eclipse is unforgettable, but make sure to protect your eyes. Looking directly at the Sun – even when it is partially covered – can cause some serious eye damage, so be prepared!

What Are the Chances of Solar Eclipses Happening?

Solar eclipses are possible thanks to a remarkable cosmic coincidence.

The Sun is about 400 times larger than the Moon – and the Moon is about 400 times closer to Earth than the Sun is. This near-perfect ratio means that the Moon can appear the same size as the Sun in our sky, allowing it to completely block the Sun's light during a total eclipse.

It's an extraordinary alignment – and one that makes a solar eclips such a rare and stunning sight.

This perfect alignment won't last forever because, as we know, nothing in space is stationary.

When Earth first gained the Moon, it orbited much closer to our planet than it does now. It has drifted roughly 1.5 inches (4 cm) away from its initial position every year.

As the Moon continues to drift outward, it will gradually appear smaller in our sky. One day – about 600 million years from now – it will be too far away to completely block the Sun's light. When that happens, total solar eclipses will become a thing of the past.

The Supermoon

An eclipse isn't the Moon's only trick – it can also appear much bigger and brighter than usual. This phenomenon is called a "supermoon" and, no, it's not just an optical illusion!

The moon doesn't orbit Earth in a perfect circle. Instead, its path is slightly elliptical, meaning the distance between the Earth and the Moon changes as it travels.

The Moon's furthest point is called an apogee and the closest point is called a perigee. When a Full Moon is within 90 per cent of its perigee point, it appears up to 14 per cent larger and about 30 per cent brighter in the sky. That's a **supermoon**!

The Blue Moon

You've probably heard the phrase "once in a blue moon" before – but it doesn't actually refer to the Moon turning blue (although it can appear bluish if smoke or dust in the atmosphere filters out red light).

A Blue Moon is the name for the second Full Moon to occur in a calendar month. Since the lunar cycle lasts about 29.5 days, two Full Moons in one month is a rare event – hence the saying!

There's also something called the Seasonal Blue Moon, which happens when a fourth Full Moon occurs in one astronomical season (winter, spring, summer or autumn/fall), instead of the usual three.

Chapter Four

**OTHER CELESTIAL BODIES
& SPACE PHENOMENA**

We've covered stars, the Sun, the planets and the Moon...

So what's left?

As it turns out, there's plenty more out there to discover – and some of it might even visible if you simply look up at the night sky!

There are comets with glowing tails, asteroids that zoom through the solar system and even galaxies beyond our own that appear as faint smudges in the night sky.

"The stars don't look bigger, but they do look brighter."

Sally Ride, the first American
woman in space, describing the view
from orbit with quiet wonder, 1998

Asteroids

When the solar system formed 4.6 billion years ago, some rocky, airless fragments were left behind. Today, we call them asteroids.

Most of these space rocks orbit the Sun in a donut-shaped region between Mars and Jupiter known as the asteroid belt. Some are huge – up to 329 miles (530 km) wide – while others are no bigger than 33 feet (10 m) in diameter.

If you want to see asteroids, you'll need some equipment – namely a telescope with an astronomy-imaging camera inside. Take a series of photos of the same sky patch over an hour, and flip through the images quickly. If you spot something moving, you may have just found an asteroid – or possibly a comet, satellite or piece of space debris.

1,422,633

The number – and counting – of known asteroids in our solar system, according to NASA's Jet Propulsion Laboratory at the California Institute of Technology.

Vesta

Remember the asteroid that we mentioned was 329 miles across? ***That's Vesta!***

The only thing larger than Vesta within the asteroid belt is the dwarf planet Ceres, so it's the largest asteroid – at least in our solar system.

Thanks to its size and highly reflective surface, Vesta is the only asteroid visible from Earth with the naked eye – when conditions are just right.

Vesta has piqued scientists' interest because it has a crust, mantle and core, much like our own planet. In 2011 – more than 200 years after Vesta was discovered – NASA sent their Dawn spacecraft to have a closer look. They found that Vesta has characteristics that bridge the gap between asteroid and planet: it has a heavily cratered surface but also contains a layered structure with a metallic core –much like a planet!

Will Vesta visit Earth?

Asteroids can depart from their typical orbits and head toward Earth – but the chances of that happening are extremely low.

As we mentioned earlier, meteors and other space debris hit the atmosphere of our planet frequently, but most of these burn up before they can reach the surface of the Earth.

NASA is one of the space-centric organizations that closely monitor larger asteroids, comets and other celestial bodies that could survive a trip through our atmosphere and pose a threat to Earth. NASA is also actively developing mitigation strategies – just in case an object ever does change course and head our way.

Comets

The leftover elements from the formation of our solar system didn't just form asteroids – they formed comets, too. While both are composed of rock and dust, comets also contain ice, setting them apart.

As a comet nears the Sun, its icy core begins to heat up. This causes a glowing ball, called the coma, to form around its solid centre, the nucleus – which is usually around 6 miles (10 km) wide. That coma can grow up to 10,000 times larger than the nucleus itself!

The heat also produces a trail of dust and gas that can stretch millions of miles, trailing behind the comet as it soars through space.

3,979

The number of comets currently identified in our solar system, according to NASA's Jet Propulsion Laboratory at the California Institute of Technology.

Halley's Comet

Halley's Comet might be the most famous comet of all – a once- (or twice-) in-a-lifetime cosmic sight!

It became noteworthy because it was the first comet proven to return to the same stretch of sky on a regular basis. This discovery showed that comets aren't one-time visitors – they orbit the Sun, just like planets do.

For the average stargazer, Halley's Comet offers a rare opportunity to see a comet with the naked eye. It appears about **once every 76 years**, making it one of the few you might witness during your lifetime.

It all began in 1705, when English astronomer **Edmond Halley** noticed something curious: a bright comet had been spotted in 1531, 1607 and 1682 – and on each occasion, the comet was given a similar description.

Drawing on Newton's theory of gravitation and planetary motions, Halley theorized that these three comets were actually one single comet returning to Earth's skies. He predicted that it would return again in 1758.

Halley's theory proved to be true – the comet reappeared just as he'd said (in December 1758, passing through perihelion – the closest point to the Sun – in 1759). Now, this famous comet carries his name as acknowledgment of his major discovery.

When Can I See Halley's Comet Next?

In the past 500 years, Halley has visited Earth, passing through perihelion, seven times on the following dates:

14 August **1531**
29 October **1607**
31 August **1682**
26 April **1759**
12 October **1835**
20 May **1910**
11 April **1986**

The last time Halley flew by Earth was in 1986. So, experts predict it will return to view from Earth in **2061**.

Are There Comets Besides Halley?

Comets mostly hang out in the outer corners of our solar system, so they're not easy to see, even with a telescope.

However, every few years, a comet tends to come close enough to Earth or the Sun that it burns bright enough for us to see. And even if we can't see the comet itself, there are signs that one has passed by. As comets burn, they shed space dust. If Earth travels through that cloud of debris and the particles burn up in our atmosphere, then we experience a **meteor shower**.

The Largest Telescope

The largest optical telescope in the world is the **Gran Telescopia Canarias (GTC)**, at the Roque de los Muchachos Observatory on Spain's Canary Islands.

This type of telescope gathers light and uses it to create a magnified image, or to collect data. The GTC relies on a 34.1-foot (10.4-m) mirror to do this part of the job.

It is used to observe a large range of astronomical phenomena, including the formation of stars and galaxies, black holes and dark matter.

2028: The Year We See Further

In 2028, the GTC is set to be dethroned by the aptly named **Extremely Large Telescope (ELT)**, built in Chile by the European Southern Observatory.

Construction began in 2017 on this colossal project, with a projected cost of around 1.3 billion euros (about £1.1 billion or US$1.36 million). Once completed, it will contain a primary mirror that stretches 128 feet (39 m) in diameter – a big leap from our current-largest telescope in Spain.

As such, the ELT will be seriously powerful. It may have the potential for observing other terrestrial planets that hover in other stars' solar systems!

Meteoroids vs Meteors vs Meteorites

Now, back to our skies and the marvels we can see from Earth – we also have meteoroids, meteors and meteorites.

It's no coincidence that these celestial bodies have very similar names!

Meteoroids are rocks that float in space. They range in size from flecks of dust to small asteroids.

When meteoroids make their way into our atmosphere and start to burn, they become **meteors**. As mentioned, meteors are also known as shooting stars.

And if a meteor survives its trip through our atmosphere – meaning it doesn't burn up completely and instead lands on the Earth's surface somewhere – it's called a **meteorite**.

60 Tonnes

That's how much the largest
intact meteorite weighed.

The massive space rock fell from the sky
around 80,000 years ago, but it was discovered
in 1920 on a farm in Namibia called Hoba West
- which is why this rock is known as the
Hoba meteorite!

Hoba is still there today, if you'd
like to go and see it!

Meteor Showers

When comets fly too close to the Sun, they burn hot and leave a trail of dust and debris in their wake. As Earth orbits the Sun, it encounters these trails from time to time. And when it does, we get a show down below!

The show is called a **meteor shower**, during which many space particles start to fall through the atmosphere at the same time. As they burn up, they streak across the night sky in a dazzling, glittering display!

As the Earth travels through the same clouds of debris in every orbit it makes, there are some meteor showers that take place every year. So, mark your calendars if you'd like to see any of the following showers:

* **December–January:** Quadrantids
* **April:** Lyrids
* **August:** Perseids
* **October:** Orionids
* **November:** Leonids
* **December:** Geminids

Fun fact: these meteor showers are named after the constellation they appear to fall from!

Perfect Conditions

Be sure to check the Moon's phase before chasing a meteor shower!

Spotting a meteor shower requires as little light pollution as possible. This includes light from a Full Moon – its brightness can make it impossible to see the show. So, check the Moon's cycle before you try to watch the shower, and choose an area where the night sky is at its darkest.

Tips for Meteor-Shower Spotting

* Find the widest stretch of open sky that you can – fields, country roads, campsites, etc

* Let your eyes adjust – they need about 20 minutes to adjust to the dark, and meteor showers come in spurts, so stick it out for a while to make sure you see the best display possible

* Don't forget comfortable accessories, like a lounge chair, hot drink and a blanket

Manmade Marvel

Step outside at at dawn or dusk and you might spot a **satellite** moving steadily across the sky. In fact, some estimates suggest that within just 15 minutes, one will cross your visual path.

As more satellites are launched into space, your chances of seeing one will continue to increase. That said, some satellites are much easier to spot than others, depending on their size, height and reflective surfaces.

9,000

The number of active satellites currently orbiting Earth (which may increase to 60,000 by 2030!).

The International Space Station (ISS)

Perhaps the easiest satellite to spot is the **International Space Station**. It's more than 300 feet (91 m) wide and it orbits Earth about 248 miles (400 km) above us!

The current home of seven astronauts from around the world, this massive structure reflects light, so it's easy to see in the sky as it gets darker.

In fact, the ISS tends to be the second-brightest thing visible in the sky at dusk, just behind the Moon itself!

So, you won't need a special telescope to see it float by – just keep your eyes peeled for one of its 16 daily orbits round our planet.

Technology Floating Through Space

* **The Hubble Space telescope**, one of the largest and most versatile telescopes launched into space

* **Progress**, a Russian cargo spaceship that brings supplies to the ISS

* **The Space X Dragon capsule**, a form of private transport into space (yes, that exists now)

The best time to spot a satellite is during summer, right after sunset or after sunrise.

Unlike other celestial bodies that are easier to see with no light, satellites are illuminated by the Sun. So, your best chance for seeing one is during dusk or dawn when there's still sunshine to reflect.

Try Your Luck With the Aurora Borealis

More commonly known as the **northern lights**, the Aurora Borealis is a breathtaking natural light show in the sky.

Solar storms, which take place on the surface of the Sun, send out clouds of electrically charged particles, and, amazingly, some of these clouds are drawn in by our planet's magnetic field and travel all the way to us.

When these particles collide with atoms and molecules in Earth's atmosphere, they heat them up – causing the atoms and molecules to glow in brilliant waves of colour that drench the sky, making for a once-in-a-lifetime viewing experience.

These stunning light displays tend to take place around the planet's magnetic poles – the North and South Pole – because they're ignited by electromagnetic particles.

Some of the best places to see the Northern Lights include:

* **Finland**

* **Sweden**

* **Norway**

* **Iceland**

* **Greenland**

* **Canada**

There are also Southern Lights (Aurora Australis), which spark at the South Pole. These are best viewed in:

* **Australia**

* **New Zealand**

* **South Africa**

* **Antarctica**

Because an Aurora Borealis or Aurora Australis sighting depend upon the electromagnetic clouds that float our way from the Sun, there's no way of predicting when they will appear, Most of the time, experts only know these displays will likely happen mere hours before they do!

2 trillion

The number of galaxies that
scientists estimate exist.

*In other words, there's
so much left to uncover!*

The Milky Way

Our solar system is just a tiny part of the Milky Way, but with the right conditions, you can see more of our galaxy stretching across the night sky.

Top tips for getting a clear view of the Milky Way:

* Pick a New Moon night, when the sky is darkest
* Head out during a clear evening with minimal cloud cover
* Find a dark spot with low light pollution – such as an empty field or park
* Give your eyes 20 minutes for your eyes to adjust to the dark

Look for the Galactic Centre, which is near the constellation Scorpio, and the Great Rift, which is a dark band that cuts vertically through the brightness of the Milky Way.

The Milky Way From...

The Northern Hemisphere

In the Northern Hemisphere, the Milky Way is best seen from June to September just after sunset. It first appears low on the east-southeast horizon, then it moves overhead and toward the west as the night goes on.

The Milky Way From...

The Southern Hemisphere

You can also spot the Milky Way with ease in the Southern Hemisphere from June to September. If you found the Sagittarius constellation and its teapot earlier, you'll have no trouble pinpointing another arm of our galaxy – you can see the Milky Way within these stars.